Median Arcuate Ligament Syndrome
Pathophysiology, Symptoms, Signs, and Management

MUHAMMAD NADEEM

The information, ideas, and suggestions in this book are not intended as a substitute
for professional medical advice. Before following any suggestions contained in
this book, you should consult your personal physician. Neither the author nor the
publisher shall be liable or responsible for any loss or damage allegedly arising as a
consequence of your use or application of any information or suggestions in this book.

ISBN: 978-1-4834-7474-8 (sc)
ISBN: 978-1-4834-7473-1 (e)

Because of the dynamic nature of the Internet, any web addresses or links contained in
this book may have changed since publication and may no longer be valid. The views
expressed in this work are solely those of the author and do not necessarily reflect the
views of the publisher, and the publisher hereby disclaims any responsibility for them.

Any people depicted in stock imagery provided by Thinkstock are models,
and such images are being used for illustrative purposes only.
Certain stock imagery © Thinkstock.

Lulu Publishing Services rev. date: 8/17/2017

Contents

Preface

In the beginning, I was writing a case report on median arcuate ligament syndrome. While writing case report, I felt there was no book available in the market on median arcuate ligament syndrome. Therefore, I decided to write book on this important topic so that doctors can get all updated information in one place. Most of the times, this important disorder is missed because it is uncommon. If books are available in the market on any rare disease, non-one can miss them. That's reason, I took step for writing book on it.

Introduction

This book focus on median arcuate ligament syndrome, which is a rare disease. This book demonstrates the pathophysiology, clinical findings, breathing effects in median arcuate ligament syndrome patients, diagnostics tests to diagnose it, surgical and non-surgical management of it, comparison among surgical procedures, complications observed in patients, and variants of median arcuate ligament syndrome. This book comprises on the ischemic complications of median arcuate ligament syndrome, psychosocial Characteristics, and outcome predictors. This book is written to compile all latest information regarding median arcuate ligament syndrome. This the first version of median arcuate ligament syndrome written by author. He will continue to upgrade the book according to the latest information.

Median Arcuate Ligament

The median arcuate ligament (MAL) is a fibrous structure of diaphragm which attaching the right and left diaphragmatic crura at T12/L1 vertebral bodies across the aortic opening. [1, 2] The crura pass anterior and superior to rim the aortic hiatus and to link the central diaphragmatic tendon. [3] Celiac ganglion is nearby the MAL and aorta is crossed by it above the celiac trunk origin. [4] It passes the truncus coeliacus higher to its ostium. The ligament insertion can be low and therefore crossover celiac axis proximal section causes a distinctive indentation. If it is located too low and thickened, it can cause compression of truncus coeliacus. [3, 5]

Median Arcuate Ligament Syndrome

The extrinsic compression of MAL, periaortic nodal tissue and prominent fibrous bands result in syndrome called median arcuate ligament syndrome (MALS), celiac artery compression syndrome (CACS) or Dunbar syndrome. [2, 6] In the small group of patients, low attachment of MAL caused compression of proximal part of the celiac artery (CA) and developed ischemic symptoms called as MALS. [2] Celiac ganglion fibers can contribute to the CA compression as well (fig.1). [7]

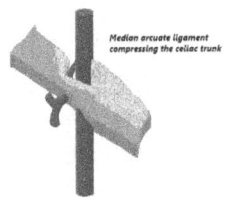

Median arcuate ligament
compressing the celiac trunk

FIGURE 1. MEDIAN ARCUATE LIGAMENT
COMPRESSING THE CELIAC ARTERY.

History

By dissecting the cadavers, Lipshutz et al elaborated it anatomically first time in 1917. He observed the overlapping of celiac artery sometimes by the diaphragmatic crura while dissecting the cadavers. [8] The mesenteric ischemia is uncommonly resulted by MALS. [9] It was first stated by Harjola in 1963 who observed mesenteric ischemia was resulted by extrinsic compression of CA and improvement of epigastric pain after decompression surgery in 57 year old man. [10] In 1965, after the surgical decompression by Daunbar et al, this vascular compression syndrome was known as clinical syndrome. [11] Thus, MALS has an anatomical and clinical characteristics. [12] It is a rare disorder [13] and the typical patients are asthenic habitus women between ages of 20 to 40 years with 3:1 female to male ratio. Young patients have higher chances, commonly thin women presenting with weight loss and epigastric pain. Mostly, the epigastric pain may be related to eating. However, it can occur in any age group. [10, 14]

Pathophysiology

Many theories have been put forward to understand the pathophysiology of its symptoms, but it is still unclear. One theory suggests that mesenteric ischemia occurs because of augmented blood flow demand through stenosis artery causing epigastric pain, while postprandial steal by collateral from superior mesenteric artery (SMA) usually avoids the foregut ischemia development causing midgut ischemia. [15, 16] Another theory explained neurogenic cause. Chronic inflammation leads to the celiac plexus overstimulation causing the splanchnic vasoconstriction with ischemia. Neural plexus morphology, location, and interconnections have variations. The fibrosis of the plexus results in decreased blood flow to the CA. [7, 17-20] Celiac plexus irritation directly and overstimulation indirectly cause pain because of subsequent splanchnic vasoconstriction and ischemia. [19, 21]

The pancreatoduodenectomy causes transient MALS with postoperatively self-resolving hepatic ischemia. Because SMA supplies the common circulation via collateral vessels confined in the pancreatic head. The pancreaticoduodenal arcades represent the collateral circulation. [22] Histoplasma capsulatum causes fibrosis elsewhere in the body leading to MALS by causing the fibrosis adjacent CA. [23] Stenosis of CA is also reported in MALS. In western countries, stenosis (87%) was reported due to arterosclerosis. [24]

The collateral arterial circulation helps in inhibiting the ischemia in single mesenteric artery stenosis. However, celiac artery compression syndrome causes mesenteric ischemia in those patients. Though there is no evidence, the collateral circulation disparity may be responsible for the alteration in clinical symptoms and variability in response to treatment. There are three groups [25]of collaterals mesenteric circulation on the basis of angiographic findings:

- Grade 0: No visible collaterals
- Grade 1: Collaterals visible on selective angiography only
- Grade 2: Collaterals seen on nonselective angiography

Celiac artery compression syndrome patients along with collateral mesenteric circulation are not much benefit from release of arcuate ligament surgery as compared to patients without the collateral circulation. The mesenteric collateral circulation classification escort and forecast shared decision making in celiac artery compression syndrome patients. [25]

Breathing Effects

The median arcuate ligament compresses the celiac axis in approximately 1% of patients which persists during inspiration causing the median arcuate ligament syndrome symptoms. [26] During the expiration, the isolated celiac axis compression may perhaps not be significant clinically. About 10-50% of healthy patients may display angiographic compression feature to a flexible limit, particularly during expiration. Mostly, these patients have no symptoms. [27] Unless it is associated with clinical findings, the only imaging finding may perhaps be insignificant and problem is neglected. Instead celiac axis compression by diaphragm MAL may cause hepatic artery thrombosis in orthotopic liver transplantation undergoing patients. [28] As stenosis of celiac results in enhancing the retrograde collateral blood flow via pancreaticoduodenal arcade from SMA to CA, a pancreaticoduodenal artery (PDA) aneurysm may happen at a low incidence rate. [29, 30]

MAL compresses the CA during respiration. The compression is more prominent during expiration in supine position person [31] because of caudally movement of diaphragm which causes visceral ischemia and postprandial abdominal pain. During inspiration and standing position, the CA slopes down to the abdominal cavity and compression is released with more vertical positioning. Some authors stated that this resulted in steal phenomenon from blood flow were turned aside from SMA through collaterals to celiac axis resulting in midgut ischemia. [32, 33]

CA run down in abdominal cavity caudally in erect position during inspiration relieving MALS symptom because of more vertical alignment of CA getting rid of compression. [34, 35] Pain mechanism is at present in discussion. Skeik et al stated that weight loss and postprandial abdominal pain in MALS were related to mesenteric ischemia. The abdominal pain is linked with undigested food regurgitation. The pathognomic characteristics of CA compression on expiration would not improve from revascularization. Therefore, he suggested that neurogenic pain and intermittent ischemia were resulted by compression of splanchnic nerve plexus and CA respectively. Both ganglion compression and changes in blood flow cause delayed gastric emptying. [36, 37]

Den et al using ultrasonic duplex scanning (UDS) of CA preoperatively stated that degree of stenosis, peak systolic blood velocity, volume of blood velocity and arterial pressure gradient in CA of 180 erect position patients during expiration, quiet breathing and intraoperatively were hemodynamically significant, but consistently lower during inspiration. [38] Ozel et al reported the elevation of peak systolic velocity during expiration, reduction of velocity during inspiration, but still remained high, and normalization in erect position. [39] Gruber et al stated that maximum expiratory peak velocity more than 350 cm/second and deflection angle >50^0 were reliable indicator for MALS. [40]

Epidemiology

MALS incidence is 1.76-4%. [41] Congenial factors are important in its etiology and are inherited. [42] Bech et al stated that it might be inherited after reporting it in monozygotic twins. [43] Congenital factors especially MAL configuration variation contribute in developing MALS. [44]

Clinical presentation

MAL causes extrinsic compression of celiac trunk in 10%-24% of patients.[10] Because of collateral supply from superior mesenteric circulation, patients are commonly asymptomatic. [10, 11] MALS presentation is variable, [45] presenting with the classical clinical triad: postprandial abdominal pain, nausea/vomiting and weight loss.[13] Those symptoms are nonspecific and are misdiagnosed as gastropathy, functional, dyspepsia or peptic ulcer disease. [37] The pain usually begins within 15-30 minutes after eating. [41] The patients can present with vague complaints including continuous, intermittent, chronic, unspecific epigastric pain getting worse after meals, or exercise, radiating to back or flanks. [4] The weight loss is associated with food fear or pain fear elicited by eating. The physical examination may activate tenderness in epigastric area or identify a bruit or thrill in the same area. [7] The pain is also accompanied with bloating and diarrhea. Sitophobia and abdominal bruit which enhanced on expiration are observed as well. [31, 46] The position of body is associated with transient symptoms improvement as leaning forward, bending the knees and bringing them near chest, because it drives MAL cephalad comparative to the CA resulting in decrease entrenchment of it on the CA. [46, 47] Renovascular hypertension caused by renal artery compression from diaphragmatic crura. [48]

Young athletes have transient abdominal pain associated with exercise. It occurs mostly because of a stitch or cramp, but MALS is a rare and ignored etiology. The typical symptoms of MALS are postprandial nausea, abdominal pain, diarrhea, and bloating, but exercise associated transient abdominal pain in young athletes is the initial presentation, which resolves with conservative or preventative approaches. If it remains in spite of these strategies, alternative reasons may be considered such as MALS. After surgical treatment of MALS, the symptoms are gone and patients return to athletics consequencely. [49]

Cusati et al reported the abdominal pain (94%), postprandial abdominal pain (80%), nausea and vomiting (55.6%), weight loss (50%), and bloating (39%) after 20 years of surgical treatment experience. [50] Epigastric bruits and tenderness are not specific for MALS. Because Bruits are identified (16%) in asymptomatic patients and younger (30%) MALS patients. [51] There is high prevalence of asymptomatic patients with radiographic proof of CA compression as Park et al [52] reported 7.3% of the asymptomatic patients had celiac axis stenosis. Other authors stated 13% to 50% asymptomatic patients had angiographic features. [53, 54] MALS may occur after Roux-en- Y gastric bypass surgery. Because Richards et al reported a MALS after bariatric surgery. [55] So, it should put into consideration if patients have postprandial pain after gastric bypass surgery.

Ischemic Complications

The compression of mesenteric arteries by MAL is commonly asymptomatic and widespread collateral vascular pathways may explain it. Nevertheless, the existence of collaterals is suggestive of ischemia development. [33] If more than one vessel is stenotic, the symptomatic risk is increased. [56] The failure of hemodynamic compensation mechanisms caused by compression (caused by extensive atherosclerotic changes later in life, or occur suddenly) lead to clinical manifestations. Hence, it is vital to recognize patients with CA compression, particularly with developed collaterals, to follow them up and treat it as a mesenteric ischemia risk factor. Mesenteric arteries compression by MAL is generally without clinical symptoms. Splendid collateral circulation in the splanchnic circulation precludes the presence of ischemic symptoms. Nevertheless, when hemodynamic compensation mechanisms stops to work, the effects can become tremendously serious. [27]

Psychosocial Characteristics

Chronic abdominal pain occurs in children and adolescents. There is 4% to 41% prevalence of chronic abdominal pain with significant direct and indirect costs to the child, family and society. MALS patients (children, adolescents) have similar psychosocial characteristics to children having other gastrointestinal illness causing chronic abdominal pain. The overlap of psychosocial and physical symptoms of MALS patients with other chronic abdominal pain illness assists to recommend that patients with chronic abdominal pain should be assessed for MALS. [57]

Outcome Predictors

There is a controversy regarding the outcome of MALS treatment. Ho et al stated that if patients had suffered from pain after exertion, MALS would respond well to decompression strategy. If they came with unprovoked pain and vomiting, they would unlikely respond the surgery. However, some studies refuse to accept the postprandial pain as a predictor of outcome. [58] The decompression procedure of median arcuate ligament syndrome shows good results in post exertional pain patients. [59] Though there is a limited guideline for median arcuate ligament syndrome management, numerous surgeons trust in subset analyses of surgical outcomes. Some surgeons suggest that less invasive surgical strategies associate with better outcomes. [60]

Diagnosis

The diagnosis of MALS is often challenging. [7] MALS is a rare disorder and typically is a diagnosis of exclusion because of mimicking symptoms with other abdominal problems. [45] Patients with MALS go through broad evaluation for ruling out other diseases comprising abdominal ultrasonography (USG), computed tomography (CT), hepatobiliary iminodiacetic acid (HIDA) scanning, and upper endoscopy. [61] If these do not succeed in identifying any other disease, MALS is assumed based on the clinical findings and work-up is shaped toward it. After clinical finding, angiography of aorta and its branches is considered as the gold standard of diagnosis. Because of low incidence, there are no guidelines for its diagnosis, treatment and follow-up. [61] Most of the median arcuate ligament syndrome patients are explored with numerus diagnostic strategies. Gastrointestinal disorder management is adopted before any certain diagnosis is accepted.

The initial screening test may be abdominal USG. [62] The USG can be performed by the laparoscopically. [63] During inspiration and expiration, abdominal USG can be used for anatomic and physiologic evaluation of CA. By showing CA moving down during expiration, causing compression and increasing velocities with post compression dilatation. [26, 36, 62, 64] The elevated peak systemic velocities during expiration undo to the normal with inspiration and standing erect. [64, 65]

Gruber et al stated that USG was the best option for diagnosis having high sensitivity (83%) and specificity (100%)compared with angiography including maximum expiratory peak velocity greater than 350 cm per second and deflection angle >50°. [40] USG shows anomalous images of SMA and CA origins, and displays the inverse flow in the hepatic artery. [19] Nevertheless, other modalities must confirm it. [7] The gold standard for diagnosis of MALS is a lateral aortic angiography.

Asymmetric focal narrowing of celiac axis with poststenotic dilation is the classic feature for diagnosis of it. [66] [7]

Respiration has effect on the narrowing by aggravating during expiration and relieving during inspiration. There is a physiological anterior and inferior aorta shift during inspiration. [67] Anterior posterior view may exhibit augmented collateral blood vessels in CA distribution territory. [7] Further noninvasive imaging studies CT angiography (CTA) and magnetic resonance angiography (MRA) help in MALS diagnosis. MRA and CTA help to diagnose the associated abdominal pathology besides MALS findings. MRA can be useful in intravenous contrast allergy patient which shows findings analogous to those of CTA (fig. 2A, 2B).

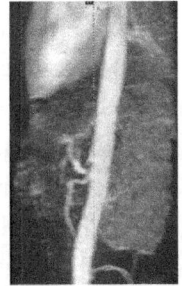

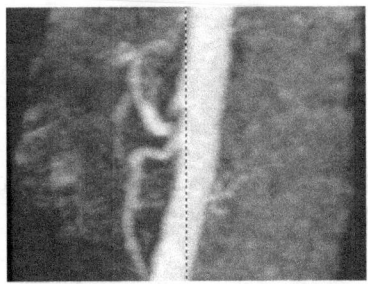

FIGURE 2A,B. MRA SHOWING THE PROXIMAL CELIAC ARTERY ABRUPT OCCLUSION IMMEDIATELY AFTER THE ORIGIN PRESUMABLY RELATED TO THE DIAGOSIS OF MEDIAC ARTCUATE LIGAMENT SYNDROME.

Lateral mesenteric angiography can be useful in diagnosing the CA stenosis or obstruction in MALS. The standard diagnosis criterion is the angiography with breathing maneuvers, there is a cephalad movement of celiac axis during inspiration causing CA compression and there is a post stenotic dilatation on expiration. Postoperatively development of recurrent symptoms can be diagnosed with angiography as well. [7, 45] The conventional angiography is used usually to diagnosis the MALS, but now three dimensional CT angiography is used to diagnose it. CT angiography joined with 3-D reconstruction can reveal all distinctive features observed with conventional angiography. Furthermore, it gives info about relationship of the CA with the diaphragm and lets compressed or stenosis artery visualization from various angles as well. [26, 68] CT angiograms show distinctive focal narrowing in the proximal celiac axis, which has a typical hooked appearance that differentiates it from other reasons of CA narrowing such as atherosclerotic disease. The celiac axis ligamentous compression leads to vascular destruction in some patients, demanding the vascular reconstruction. [26]

MALS leads to gastric ischemia which is diagnosed by gastric exercise established in limited studies to be an effective way for diagnosis and follow up evaluation as well. It is an adjunctive modality in which raised intraluminal and intramucosal $PaCO_2$ levels propose gastric ischemia. $PaCO_2$ levels are counted before, during and after ten minutes of submaximal exercise. The gastric arterial $PaCO_2$ difference more than 0.8 kPa after exercise, the arterial lactate level < 72 mg/dl and the rise in gastric $PaCO_2$ level after exercise are the indications for pathological findings. [7, 45, 69]

Even though imaging and gastric exercise tonometry modalities aid to diagnose MALS patients, percutaneous celiac ganglion block can assist in diagnosing the patients who respond well to surgical strategy. Thus, it helped to recognize the patients preoperatively who would get benefit from MALS release. [20, 47] The concept of this modality is that the inflammation and compression of celiac plexus, relay center for abdominal visceral afferent fibers carrying pain sensation, cause the MALS symptoms. The percutaneous injection of the celiac ganglion with anesthetic agents like bupivacaine, lidocaine etc. results in momentary relief and with ethanol results in permanent relief of symptoms. The celiac ganglion block has conventionally been utilized for the pigheaded pain relief characteristically allied to untreatable malignant and benign disease. [45, 70]

4-D wide area CT angiogram

The four dimensional (4D) wide area CT angiogram is a novel application for investigation of MALS. It is a non-invasive strategy to obtain high-resolution images of both coeliac axis and the MAL. It elaborates the relationship of MAL and coeliac axis with each other during respiration. It helps in surgical planning by identifying post-stenotic dilatation of the artery and collateral vessels and determining the hemodynamic significance of the arterial narrowing. In addition, it assesses the other reasons of arterial narrowing, for example atherosclerosis and extrinsic compression. [71] It can be a useful initial investigation for coeliac axis compression in conjunction with other reported strategies for example gastric exercise tonometry and superior mesenteric artery angiogram vasodilator injection. [69, 72] It competes the conventional catheter angiography at reduced radiation exposure. [71]

Management

MALS treatment purpose is to restore normal blood flow in the celiac axis and abolish neural irritation developed by the celiac ganglion fibers. The first published case accomplished this by eliminating the celiac plexus constrictive fibers and dividing the MAL. [10, 11] The long term results do not prove satisfactory. Still some investigators questioned the MALS existence as a clinical entity and proposed that same results would be acquired with non-surgical modalities. [53]

In 1985, Reilly et al [73] defined patient criteria correlated with a successful surgical treatment, comprising age (40-60 years old), female, postprandial abdominal pain, weight loss (greater than 20lb), and an angiogram with augmented collateral flow or poststenotic dilatation. In addition, MAL release alone is not enough in the treatment of it, additional surgical management should be considered as well like graduated celiac dilatation or reconstruction. There are no significant differences in results between these two surgical strategies. There are different sorts of vascular reconstruction comprising patchy CA angioplasty, CA aortic reimplantation, and aortoceliac bypass with Dacron graft or saphenous vein. [7] Many types of treatment modalities are endorsed. The vascular reconstruction along with MAL release is suggested in CA stenosis situation or palpable thrill condition. [74]

The assessment of celiac blood flow before and after MAL release is either pressure gradient measurement in the aorta and celiac (splenic) artery using angiography probe or measuring the velocity in the CA using doppler ultrasound. If the pressure gradient between the aorta and CA is 10 mm Hg, it will be significant for presence of the CA stenosis. If velocity measurement via Doppler ultrasound or pressure gradient measurement via angiography show compromised blood flow, the patient will benefit from endoluminal strategies such as balloon angioplasty with or without stent deployment in the CA, before undergoing the vascular reconstruction procedure which associates with increased morbidity. [7]

Percutaneous transluminal angioplasty or endovascular treatment with or without stenting alone does not benefit in the MALS treatment. The persistent MAL compression on CA causes extrinsic luminal constricting resulting in perpetual modifications in the vessel wall. The long term success rate in patents undergone decompression alone and decompression along with revascularization afterward is 53% and 76% respectively (fig. 3, 4). So, release of MAL and vascular reconstruction is undertaken afterward. [15, 73-76]

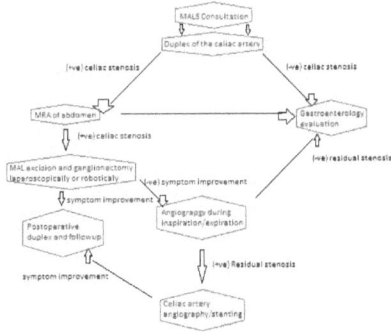

FIGURE 3. MEDIAN ARCUATE LIGAMENT SYNDROM (MALS) TREATMENT ALGORITHM, MAGNETIC RESONANCE ANGIOGRAPHY (MRA). [77]

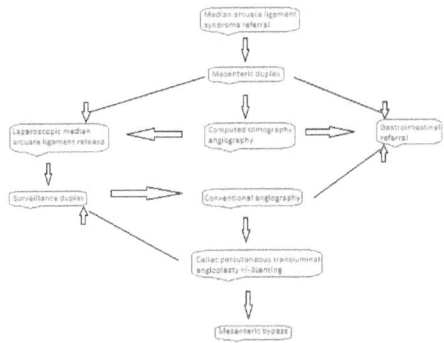

FIGURE 4. MULTIDISCIPLINARY CLINICAL DIAGNOSTIC AND THERAPEUTIC ALGORITHM. [78]

Nonsurgical Treatment

CA dissection without accompanied aortic dissection is a rare condition. It is related to MALS as well. It can be treated conservatively in stable patient, though patients having retroperitoneal hemorrhage. Long-term antiplatelet or anticoagulation treatment is not necessary. [79]

Surgical Treatment

Laparoscopic surgical approach is the most noteworthy surgical advances of the 20th century and eminent in numerous surgical techniques. [80-82] The ideal method of relieving symptomatic compression is the laparoscopic technique. [83] In 2000, laparoscopic treatment of MALS was first introduced.[63] Celiac artery reconstruction is often needed, because extrinsic compression of celiac artery usually causes chronic intrinsic stenosis, short occlusion or dissection of the celiac artery. It is typically done with the open strategy. [47]

Open Technique

The open surgical approach is achieved by dissection of the diaphragm crus proximal to the celiac artery and the MAL. Neurolysis and ganglionic tissue over the aorta dissection is executed after the open procedure. [47, 73] A short upper midline laparotomy is performed in the open approach. The Laparoscopic approach is performed by dissection of the diaphragm crus proximal to the celiac artery and the MAL. Metal dilators are got in through transverse cut made in the splenic artery, or through short longitudinal or transverse cut made in the distal, commonly dilated celiac artery. This strategy occasionally works, as it is critical to leave behind a celiac artery with a dissected or fragmented intima, caused by the process of metal dilator. Reconstruction can be achieved with a patch or more often with a large 8 mm shorter polytetrafluoroethylene or polyester interposition grafts. If stents are positioned previously, those have to be taken away and an interposition graft is used for artery reconstruction. [4, 47]

Laparoscopic Technique

The laparoscopic approach is accomplished by cutting the ganglionic tissue over the aorta. The fibrous stenosis, occlusion or aneurysmal degeneration demand the celiac artery revascularization. The median blood loss in laparoscopic strategy is 5 ml and median operative time is 118 minutes. [78]

The laparoscopic technique is performed under general anesthesia. Patient is placed in supine posture. In standardized procedure, patient's legs are split and surgeon stands in between them. Abdominal pressure is maintained at 14 mm of Hg by filling the CO_2 in the abdomen, which expands the abdomen making the surgical field clear. A 10 mm incision is made at the umbilicus for camera port. A 5 mm epigastric port placed. A liver retractor is inserted through epigastric port for lifting the liver left lobe. Two 5 mm cannulas are inserted in the left and right upper abdomen. There is an additional 5 mm cannula inserted in the left abdomen. Patient is placed in a reverse Trendelenburg posture with right or left tilt as needed. 30^0 scope is utilized for better visualization of the gastroesophageal junction. [84]

First, pars flaccida is dissected, and both crura are incised then. Along with penrose drain, the esophagus is looped. Then diaphragmatic right crus is skeletonized. After it, abdominal aorta is exposed and marked out to pinpoint the celiac axis origin. Left hepatic and gastric arteries incision assist in pinpointing the celiac axis origin on the aorta as well. For getting better vision, stomach is retracted to the left. Abnormally inserted median arcuate ligament fibers are identified near the celiac axis origin. Wholly fibro-fatty tissue on either side of celiac axis origin is detached to bare the origin of celiac axis fully. When these maneuvers are accomplished, the celiac axis is clearly envisioned without any remaining kinking. [84]

Robotic Assisted Technique

The introduction of robotic surgery has assisted many minimally invasive surgeries and got better the results. [85, 86] The robotic assisted surgery, also called as computer assisted surgery or surgical telemanipulation, has been developed to overwhelmed human restrictions and abolish barriers related to conventional surgical and interventional apparatuses. [85-89] The introduction of this technology may enhance dexterity and accuracy in execution the difficult tasks and decrease learning times in the challenging scope of the abdominal vascular surgery. [90] Robotic assisted laparoscopic approach has been achieved in a few MALS patients. [77] It makes better visualization and dexterity compared with laparoscopic approach. [91]

The simple division of celiac plexus and MAL is not a good strategy for all patients. [92] The celiac decompression has 53% of the long term success rate, but celiac decompression followed by celiac revascularization has 76% of the success rate. The presence of persistent vessel deformity, pressure gradient or thrill after the decompression procedure are indications of reconstruction approach. [73] The comprehensive celiac artery skeletonization may not show suitable long term symptoms improvement in some patients. Thus, any procedure accomplished for celiac artery compression treatment requisites the artery assessment after decompression for flow adequacy. [74]

Transthoracic Approach

Median arcuate ligament can be released easily and safely through low anterolateral thoracotomy. Previous supramesocolic surgeries, presence of abdominal wall hernias, stomas, pancreatitis, previous intra-abdominal sepsis, or radiation treatment make the transperitoneal approach unsafe in mesenteric revascularization. Patients have calcified abdominal aortoiliac fragment along with visceral artery occlusive disease, which causes inadequate arterial inflow source. In those circumstances, peritoneal cavity is avoided and proximal to the abdominal aorta arterial inflow source is important. SMA and celiac artery can be exposed for bypass from the instantly proximal aorta. This strategy is suitable for median arcuate ligament syndrome patients. It allows sufficient ligament transection and autonomic splanchnic ganglia resection. [93]

The thoracoabdominal route to the descending distal thoracic aorta and the visceral aortic segment allows concurrent proximal visceral arteries and distal descending thoracic aorta exposure, which allows the short bypasses construction, leading to perfect long term graft patency rates along with significant morbidity. Without the need for the more extensive and potentially more morbid thoracoabdominal incision, the SMA and celiac artery are exposed through a low thoracotomy. This approach is less extensive than thoracoabdominal, retroperitoneal perivisceral and supraceliac aorta exposure. [93]

Comparison among surgical procedures

The laparoscopic approach compared to open approach gives good visibility and exposure. The open approach permits the direct reconstruction such as aorto-coeliac bypass or patch, while the laparoscopic approach needs an extra endovascular procedure with the celiac artery stenting. [4, 47] Percutaneous transluminal angioplasty with celiac artery stenting is performed for recurrent symptoms after laparoscopic release. Angioplasty is not executed alone. [78] The laparoscopic Doppler ultrasound scanning provides information about degree of compression and flow adequacy after dissection of the compression fibers. [63]

Therefore, it is the vital part of the surgery. The percutaneous angioplasty and stenting of the celiac artery are the vital revascularization techniques during laparoscopic approach. However, the angioplasty and stenting are applied in the atherosclerotic disease treatment, their application in celiac artery compression are debatable, because extrinsic compression inhibits sufficient vessel dilatation. [94-96] The use of intracorporeal suturing in the laparoscopic revascularization is another available choice. [97] In short, Laparoscopic technique leads the surgeon to skeletonize the celiac axis even though getting rid of a laparotomy, which causes the short hospital stay, reduced postoperative pain, and a quick recovery. It limits the intraoperative blood loss and postoperative adhesions as well. [98]

Surgeons should be careful about the over diagnosis of celiac artery compression syndrome. They should be cautioned not to permit the less invasive nature of this approach sway them toward a lenient selection of patients and unwarranted profuse use of this approach. [63] Endovascular treatment is not recommended for celiac artery stenting if prior external compression release of celiac artery is not treated. If it is done without external compression release, there will be the risk of stent crush or rupture. [47]

Follow up

Table 1 is demonstrating the incidence of recurrence of symptoms in follow-up of the patients with median arcuate ligament syndrome. Mostly patients got immediately symptoms relief. Very few patients need for conversion to open decompression because of complications. No surgery related fatalities are reported so far (table 1).

TABLE 1. DISPLAYING THE CASES WITH LAPAROSCOPIC TREATMENT OF MEDIAN ARCUATE LIGAMENT SYNDROME ALONG WITH IMPROVEMENT OF SYMPTOMS, MEAN FOLLOW UP, AND RECURRENCE OF SYMPTOMS.

Author, year	No. of cases	Improvement of symptoms	Mean follow-up duration	Recurrence of symptoms
Van Petersen, 2009[99]	42	41	20 months	Not mentioned
Baccari, 2009[100]	16	14	28 months	0
Roseborough, 2009[101]	15	14	44 months	1
Tulloch, 2010[102]	12	12	14 months	5

Author, year	No. of cases	Improvement of symptoms	Mean follow-up duration	Recurrence of symptoms
Rotellar, 2010[32]	07	03	06 months to 8yrs	Not mentioned
Aschenbach, 2011[103]	22	22	Not mentioned	Not mentioned
Berard, 2012[104]	11	10	35 months	1
Nguyen, 2012 [105]	05	05	20 months	0
Do, 2013 [106]	12	08	22 months	Not mentioned
El-Hayek, 2013[107]	15	14	15 months	1
Joyce, 2014 [108]	06	06	13 months	0

Median arcuate ligament syndrome is the foundation of pathological hemodynamic variations in the abdominal vasculature. Development of collateral circulation emphasizes the follow-up visits.

Complications

The incidences of major postoperative complications are 6.5% in the patients undergone open procedure. The most common complication is thrombosed bypass graft (2%). Further complications are stroke (1.4%), pancreatitis (1%), gastroesophageal reflux disease (1%), splenic infarction (0.3%) and hemothorax (0.3%). No procedure related deaths are reported so far in any of the open series. [77]

The intraoperative complications in the laparoscopic procedure are bleeding from visceral artery (4.1%), pneumothorax (2.5%), bleeding from aorta (1.7%), and laceration of phrenic artery (0.8%) (table 2). Gastroparesis (0.8%) and pancreatitis (0.8%) are the postoperative complication observed after laparoscopic procedure. No procedure related fatalities are observed so far in any of the laparoscopic series. [77] MALS causes the visceral artery aneurysms which should be treated by endovascular strategy coupled with celiac artery decompression method to restore physiologic blood flow, when size ratio approaches three. [109]

TABLE. 02. INTRAOPERATIVE COMPLICATIONS DURING LAPAROSCOPIC DECOMPRESSION OF CELIAC ARTERY FOR MEDIAN ARCUATE LIGAMENT SYNDROM. [77]

Intraoperative complications	Laparoscopic surgery
Bleeding from celiac artery	05
Pneumothorax	03
Bleeding from suprarenal artery	01
Bleeding from gastric artery	01
Laceration of phrenic artery	01
Aortic puncture	01

Variants and Associations

Spinal Deformity correction

The CA compression can occur in extensive correction of sagittal balance in spinal deformity adult. Notani et al reported a first case of 77 year-old female who was undergone a two stage correction. Her lumbar lordosis got better from -47^0 to 53^0. After the procedure, she had diarrhea, vomiting. She developed MALS diagnosed with CT, which was treated surgically. Thus, Spinal surgeons must be aware of the it in treating the spinal kyphosis deformity. [110]

Gastroduodenal artery (GDA) and pancreaticoduodenal artery (PDA) aneurysms

MALS is often allied to prominence of the pancreaticoduodenal arcade, a collateral circulation between the celiac trunk and SMA.[111] GDA and PDA aneurysms are rare lesions and are defined as the lesions are 1.5 times the size of the normal vessel on imaging. First time in 1973, Sutton and Lawton demonstrated a PDA aneurysm related to celiac axis constriction.[112] GDA and PDA aneurysms are accompanied with significant rupture risk. These are allied to celiac axis occlusion or stenosis causing a changed hemodynamics in the pancreaticoduodenal arcade. [113] MAL is the most common reported reason of PDA aneurysm by keeping pressure on it. PDA aneurysm related to CA occlusion as a result of aortic syndrome is exceptionally rare. [114]

MAL compresses the celiac axis resulting in superior indentation presented a few millimeters from the vessel origin, often causing post-stenotic dilation. Most patients experienced abdominal pain, nausea/vomiting and diagnosed with gastrointestinal hemorrhage. These are inclined to rupture irrespective of their size. Thus management is accordingly begun for all aneurysms at identification. Endovascular repair is considered as the first line treatment for maintaining the hepatic circulation and celiac axis obstruction. [113] The GDA and its superior PDA branches, and inferior PDA branches of SMA are the part of the pancreaticoduodenal arcade. These account for 2% of all visceral artery aneurysms. [115] They rupture at an uneven rate compared with other visceral aneurysms and associate with increased rate of morbidity and mortality. [116] The most peripancreatic aneurysms are false, associated with infection, trauma, or inflammation. [115] True aneurysms are rare lesions that are generally associated with augmented retrograde blood flow through the arcade in the presence of a proximal celiac axis obstruction. [117]

These lesions are treated by revascularizing the celiac axis along with or without MAL detachment, but mostly managed traditionally with open suture tying or aneurysmectomy. With the advance in surgical interventions, the lesions can be managed by coil embolization along with or without celiac axis stenting. The Short term durability of endovascular technique is successful but to assess the long term durability demands future studies. Thus endovascular technique is considering first line therapy for pancreadtico duodenal arcade aneurysms in which hepatic arterial flow should be preserved. After axial imaging study preoperatively, patient's arterial anatomy should be reviewed thoroughly. Hepatic artery flow must be maintained for coil embolization of GDA aneurysms in celiac axis occlusion or stenosis. Celiac axis stenting or aortohepatic artery bypass keeps flow through the occluded celiac axis for maintaining hepatic artery flow in GDA coil embolization. The revascularization of celiac is not compulsory in the existence of a replaced right hepatic artery originating from the SMA. If endovascular repair is not possible, open repair technique must be used. PDA aneurysms do not need to coil embolize the celiac axis revascularization, as patent GDA keeps hepatic artery flow. [113]

Arc of Buhler (AOB)

The AOB was first explained by Buhler and Tandler. It is developed in prenatal condition by the recession failure of the 10th and 13th ventral segmental aorta arteries resulted in a pertinacious ventral anastomosis between the SMA and CA embryonically. [118, 119] The recession failure causes the tenacious arterial linking resulted in AOB. It is the free of the pancreaticoduodenal and dorsal pancreatic arteries, those are common anastomosis between CA and SMA or one of their branches.[120] It is a rare anomaly having incidence 1-4% in symptomatic mesenteric arterial stenosis patients undergoing an arteriography, and 3% in asymptomatic liver transplant donor undergoing mapping arteriography. [120-122] The AOB is generally identified incidentally in asymptomatic patients who go through imaging for other medical conditions or in symptomatic patients having celiac trunk stenosis regarded to arterial sclerosis or MALS. Its aneurysms are rarer. The rupture of those aneurysms is a life threatening condition. The diameter of aneurysms about 4-6 cm have been reported. Though their management still remain unclear. But they are managed surgically by transcatheter arterial embolization. [123]

Reference

1   Ghulam, Q.M., et al., *[Median arcuate ligament syndrome]*. Ugeskr Laeger, 2015. **177**(39): p. V03150284.

2   Karavelioglu, Y., M. Kalcik, and T. Sarak, *Dunbar syndrome as an unusual cause of exercise-induced retrosternal pain*. Turk Kardiyol Dern Ars, 2015. **43**(5): p. 465-7.

3   Ng, F.H., et al., *Median arcuate ligament syndrome*. Hong Kong Med J, 2016. **22**(2): p. 184.e3-4.

4   Czihal, M., et al., *Vascular compression syndromes*. Vasa, 2015. **44**(6): p. 419-34.

5   Duckheim, M., et al., *A patient presenting with stress-induced epigastric pain*. BMJ Case Rep, 2015. **2015**.

6   Romero, M.E., et al., *[Celiac trunk compression syndrome by the median arcuate ligament. Laparoscopic approach]*. Medicina (B Aires), 2015. **75**(3): p. 169-72.

7   Duffy, A.J., et al., *Management of median arcuate ligament syndrome: a new paradigm*. Ann Vasc Surg, 2009. **23**(6): p. 778-84.

8   Lipshutz, B., *A COMPOSITE STUDY OF THE COELIAC AXIS ARTERY*. Ann Surg, 1917. **65**(2): p. 159-69.

9   Biyikoglu, I., et al., *Crohn's disease masked by median arcuate ligament syndrome*. Chin Med J (Engl), 2013. **126**(14): p. 2798.

10 Harjola, P.T., *A RARE OBSTRUCTION OF THE COELIAC ARTERY. REPORT OF A CASE*. Ann Chir Gynaecol Fenn, 1963. **52**: p. 547-50.

11 Dunbar, J.D., et al., *Compression of the celiac trunk and abdominal angina*. Am J Roentgenol Radium Ther Nucl Med, 1965. **95**(3): p. 731-44.

12 Lindner, H.H. and E. Kemprud, *A clinicoanatomical study of the arcuate ligament of the diaphragm*. Arch Surg, 1971. **103**(5): p. 600-5.

13 Kotarac, M., et al., *Surgical treatment of median arcuate ligament syndrome: case report and review of literature*. Srp Arh Celok Lek, 2015. **143**(1-2): p. 74-8.

14 Sultan, S., et al., *Eight years experience in the management of median arcuate ligament syndrome by decompression, celiac ganglion sympathectomy, and selective revascularization*. Vasc Endovascular Surg, 2013. **47**(8): p. 614-9.

15 Cina, C.S. and H. Safar, *Successful treatment of recurrent celiac axis compression syndrome. A case report*. Panminerva Med, 2002. **44**(1): p. 69-72.

16 Loukas, M., et al., *Clinical anatomy of celiac artery compression syndrome: a review.* Clin Anat, 2007. **20**(6): p. 612-7.

17 Rob, C., *Stenosis and thrombosis of the celiac and mesenteric arteries.* Am J Surg, 1967. **114**(3): p. 363-7.

18 Gutnik, L.M., *Celiac artery compression syndrome.* Am J Med, 1984. **76**(2): p. 334-6.

19 Bech, F.R., *Celiac artery compression syndromes.* Surg Clin North Am, 1997. **77**(2): p. 409-24.

20 Duncan, A.A., *Median arcuate ligament syndrome.* Curr Treat Options Cardiovasc Med, 2008. **10**(2): p. 112-6.

21 Balaban, D.H., et al., *Median arcuate ligament syndrome: a possible cause of idiopathic gastroparesis.* Am J Gastroenterol, 1997. **92**(3): p. 519-23.

22 Sanchez, A.M., et al., *Temporary medium arcuate ligament syndrome after pancreatoduodenectomy.* Am Surg, 2013. **79**(2): p. E58-60.

23 Niemann, N., F.L. Hochman, and R.S. Huang, *Histoplasmosis as a possible cause of retroperitoneal fibrosis and median arcuate ligament syndrome: A case report.* Int J Surg Case Rep, 2014. **5**(8): p. 473-5.

24 Berney, T., et al., *The role of revascularization in celiac occlusion and pancreatoduodenectomy.* Am J Surg, 1998. **176**(4): p. 352-6.

25 van Petersen, A.S., et al., *Clinical significance of mesenteric arterial collateral circulation in patients with celiac artery compression syndrome.* J Vasc Surg, 2017. **65**(5): p. 1366-1374.

26 Horton, K.M., M.A. Talamini, and E.K. Fishman, *Median arcuate ligament syndrome: evaluation with CT angiography.* Radiographics, 2005. **25**(5): p. 1177-82.

27 Arazinska, A., et al., *Median arcuate ligament syndrome: Predictor of ischemic complications?* Clin Anat, 2016. **29**(8): p. 1025-1030.

28 Jurim, O., et al., *Celiac compression syndrome and liver transplantation.* Ann Surg, 1993. **218**(1): p. 10-2.

29 Matsumura, Y., et al., *Median arcuate ligament syndrome presenting as hemorrhagic shock.* The American Journal of Emergency Medicine, 2013. **31**(7): p. 1152.e1-1152.e4.

30 Takase, A., et al., *Two patients with ruptured posterior inferior pancreaticoduodenal artery aneurysms associated with compression of the celiac axis by the median arcuate ligament.* Ann Vasc Dis, 2014. **7**(1): p. 87-92.

31 Bagley, J., et al., *Case report: median arcuate ligament syndrome diagnosed with computed tomography and Doppler ultrasonography.* Radiol Technol, 2015. **86**(3): p. 238-45.

32 J, A.C., et al., *The celiac axis compression syndrome (CACS): critical review in the laparoscopic era.* Rev Esp Enferm Dig, 2010. **102**(3): p. 193-201.

33 Arazinska, A., et al., *An unusual case of left renal artery compression: a rare type of median arcuate ligament syndrome.* Surg Radiol Anat, 2016. **38**(3): p. 379-82.

34 Kim, S.J., et al., *Open surgical decompression of celiac axis compression by division of the median arcuate ligament.* J Korean Surg Soc, 2013. **85**(2): p. 93-5.

35 Chou SQ, K.K., Wong LS, Fung DH,, *Imaging features of median arcuate ligament syndrome.* J Hong Kong Col Radiol, 2010. **13**.

36 Skeik, N., et al., *Median arcuate ligament syndrome: a nonvascular, vascular diagnosis.* Vasc Endovascular Surg, 2011. **45**(5): p. 433-7.

37 Gunduz, Y., et al., *Clinical and radiologic review of uncommon cause of profound iron deficiency anemia: median arcuate ligament syndrome.* Korean J Radiol, 2014. **15**(4): p. 439-42.

38 Den, B., et al., *[The significance of respiratory and orthostatic tests in duplex scanning in diagnostics of celiac artery compression syndrome].* Vestn Khir Im I I Grek, 2013. **172**(2): p. 28-31.

39 Ozel, A., et al., *Ultrasonographic diagnosis of median arcuate ligament syndrome: a report of two cases.* Med Ultrason, 2012. **14**(2): p. 154-7.

40 Gruber, H., et al., *Ultrasound of the median arcuate ligament syndrome: a new approach to diagnosis.* Med Ultrason, 2012. **14**(1): p. 5-9.

41 Göya C, H.C., Hattapoğlu S, Çetinçakmak MG, Teke M, Kuday S, *Diagnosis of median arcuate ligament syndrome on multidetector computed tomography.* J Med Cases, 2013. **4**(9).

42 Okten, R.S., et al., *Is celiac artery compression syndrome genetically inherited?: a case series from a family and review of the literature.* Eur J Radiol, 2012. **81**(6): p. 1089-93.

43 Bech, F., et al., *Median arcuate ligament compression syndrome in monozygotic twins.* J Vasc Surg, 1994. **19**(5): p. 934-8.

44 Ilica, A.T., et al., *Median arcuate ligament syndrome: multidetector computed tomography findings.* J Comput Assist Tomogr, 2007. **31**(5): p. 728-31.

45 Kim, E.N., et al., *Median Arcuate Ligament Syndrome-Review of This Rare Disease.* JAMA Surg, 2016. **151**(5): p. 471-7.

46 Lainez, R.A. and W.S. Richardson, *Median arcuate ligament syndrome: a case report.* Ochsner J, 2013. **13**(4): p. 561-4.

47 Gloviczki, P. and A.A. Duncan, *Treatment of celiac artery compression syndrome: does it really exist?* Perspect Vasc Surg Endovasc Ther, 2007. **19**(3): p. 259-63.

48 Bunger, C.M., et al., *Laparoscopic treatment of renal artery entrapment.* J Vasc Surg, 2010. **52**(5): p. 1357-61.

49 Harr, J.N., I.N. Haskins, and F. Brody, *Median arcuate ligament syndrome in athletes.* Surg Endosc, 2016.

50 Cusati DA, N.A., Gloviczki P, *Median arcuate ligament syndrome: 20 -year experience of surgical treatment,* in *60th Annual Meeting of the Society for Vascular Surgery.* 2006: Philadelphia, PA.

51 Julius, S. and B.H. Stewart, *Diagnostic significance of abdominal murmurs.* N Engl J Med, 1967. **276**(21): p. 1175-8.

52 Park, C.M., et al., *Celiac axis stenosis: incidence and etiologies in asymptomatic individuals.* Korean J Radiol, 2001. **2**(1): p. 8-13.

53 Szilagyi, D.E., et al., *The celiac artery compression syndrome: does it exist?* Surgery, 1972. **72**(6): p. 849-63.

54 Bron, K.M. and H.C. Redman, *Splanchnic artery stenosis and occlusion. Incidence; arteriographic and clinical manifestations.* Radiology, 1969. **92**(2): p. 323-8.

55 Richards, N.G., et al., *Celiac artery compression after a gastric bypass.* Surg Laparosc Endosc Percutan Tech, 2014. **24**(2): p. e66-9.

56 Thomas, J.H., et al., *The clinical course of asymptomatic mesenteric arterial stenosis.* J Vasc Surg, 1998. **27**(5): p. 840-4.

57 Mak, G.Z., et al., *Pediatric Chronic Abdominal Pain and Median Arcuate Ligament Syndrome: A Review and Psychosocial Comparison.* Pediatr Ann, 2016. **45**(7): p. e257-64.

58 Ho, K.K., et al., *Outcome predictors in median arcuate ligament syndrome.* J Vasc Surg, 2017.

59 Ho, K.K.F., et al., *Outcome predictors in median arcuate ligament syndrome.* J Vasc Surg, 2017. **65**(6): p. 1745-1752.

60 Grant, Y., et al., *Exercise induced median arcuate ligament syndrome in athletes: case series.* J Sports Med Phys Fitness, 2017.

61 Celik, S., et al., *A case of pancreatic cancer with concomitant median arcuate ligament syndrome treated successfully using an allograft arterial transposition.* J Surg Case Rep, 2015. **2015**(12).

62 Scholbach, T., *Celiac artery compression syndrome in children, adolescents, and young adults: clinical and color duplex sonographic features in a series of 59 cases.* J Ultrasound Med, 2006. **25**(3): p. 299-305.

63 Roayaie, S., et al., *Laparoscopic release of celiac artery compression syndrome facilitated by laparoscopic ultrasound scanning to confirm restoration of flow.* J Vasc Surg, 2000. **32**(4): p. 814-7.

64 Wolfman, D., E.I. Bluth, and J. Sossaman, *Median arcuate ligament syndrome.* J Ultrasound Med, 2003. **22**(12): p. 1377-80.

65 Erden, A., M. Yurdakul, and T. Cumhur, *Marked increase in flow velocities during deep expiration: A duplex Doppler sign of celiac artery compression syndrome.* Cardiovasc Intervent Radiol, 1999. **22**(4): p. 331-2.

66 Rubinkiewicz, M., et al., *Laparoscopic decompression as treatment for median arcuate ligament syndrome.* Ann R Coll Surg Engl, 2015. **97**(6): p. e96-9.

67 Tracci, M.C., *Median arcuate ligament compression of the mesenteric vasculature.* Tech Vasc Interv Radiol, 2015. **18**(1): p. 43-50.

68 Kopecky, K.K., et al., *Median arcuate ligament syndrome with multivessel involvement: diagnosis with spiral CT angiography.* Abdom Imaging, 1997. **22**(3): p. 318-20.

69 Mensink, P.B., et al., *Gastric exercise tonometry: the key investigation in patients with suspected celiac artery compression syndrome.* J Vasc Surg, 2006. **44**(2): p. 277-81.

70 Lee, M.J., et al., *CT-guided celiac ganglion block with alcohol.* AJR Am J Roentgenol, 1993. **161**(3): p. 633-6.

71 Cheng, K., et al., *Novel application of four-dimensional wide-area detector computed tomographic angiography for investigation of median arcuate ligament syndrome.* J Med Imaging Radiat Oncol, 2017. **61**(2): p. 239-242.

72 Kalapatapu, V.R., et al., *Definitive test to diagnose median arcuate ligament syndrome: injection of vasodilator during angiography.* Vasc Endovascular Surg, 2009. **43**(1): p. 46-50.

73 Reilly, L.M., et al., *Late results following operative repair for celiac artery compression syndrome.* J Vasc Surg, 1985. **2**(1): p. 79-91.

74 Takach, T.J., et al., *Celiac compression syndrome: tailored therapy based on intraoperative findings.* J Am Coll Surg, 1996. **183**(6): p. 606-10.

75 Delis, K.T., et al., *Median arcuate ligament syndrome: open celiac artery reconstruction and ligament division after endovascular failure.* J Vasc Surg, 2007. **46**(4): p. 799-802.

76 Wang, X., et al., *Celiac revascularization as a requisite for treating the median arcuate ligament syndrome.* Ann Vasc Surg, 2008. **22**(4): p. 571-4.

77 Jimenez, J.C., M. Harlander-Locke, and E.P. Dutson, *Open and laparoscopic treatment of median arcuate ligament syndrome.* J Vasc Surg, 2012. **56**(3): p. 869-73.

78 Columbo, J.A., et al., *Contemporary management of median arcuate ligament syndrome provides early symptom improvement.* J Vasc Surg, 2015. **62**(1): p. 151-6.

79 Hosaka, A., M. Nemoto, and T. Miyata, *Outcomes of conservative management of spontaneous celiac artery dissection.* J Vasc Surg, 2016.

80 Nadeem.M, K.S., Ali.S, Shafiq.M, Elahi.MW, Abdullah.F, Hussain.I., *Comparison of extra-corporeal knot-tying suture and metallic endo-clips in laparoscopic appendiceal stump closure in uncomplicated acute appendicitis.* International Journal of Surgery Open, 2016. 2: p. 11-14.

81 Muhammad Nadeem, Z.A.C., Burhan Ul haq, *Comparison between Single Incision Laparoscopic Surgery and Conventional Three Port Laparoscopic Cholecystectomy.* BAOJ Surgery, 2016. 02(02): p. 012.

82 Nadeem, M., *Stretta; Treatment for Gastro Esophageal Disease.* BAOJ Obe Weigt Manage, 2016. 2(1): p. 008.

83 El Asmar, A. and Z. El Rassi, *Median arcuate ligament syndrome: case presentation and video-illustrated laparoscopic management.* Clin Case Rep, 2016. 4(12): p. 1213-1214.

84 Ramakrishnan, P., et al., *Laparoscopic Division of Median Arcuate Ligament for the Celiac Axis Compression Syndrome-Two Case Reports with Review of Literature.* Indian J Surg, 2016. 78(2): p. 163-5.

85 Howe, R.D. and Y. Matsuoka, *Robotics for surgery.* Annu Rev Biomed Eng, 1999. 1: p. 211-40.

86 Lanfranco, A.R., et al., *Robotic surgery: a current perspective.* Ann Surg, 2004. 239(1): p. 14-21.

87 Camarillo, D.B., T.M. Krummel, and J.K. Salisbury, Jr., *Robotic technology in surgery: past, present, and future.* Am J Surg, 2004. 188(4A Suppl): p. 2s-15s.

88 Ruurda, J.P., T.J. van Vroonhoven, and I.A. Broeders, *Robot-assisted surgical systems: a new era in laparoscopic surgery.* Ann R Coll Surg Engl, 2002. 84(4): p. 223-6.

89 Hanly, E.J. and M.A. Talamini, *Robotic abdominal surgery.* Am J Surg, 2004. 188(4A Suppl): p. 19s-26s.

90 Martinez, B.D. and C.S. Wiegand, *Robotics in vascular surgery.* Am J Surg, 2004. 188(4A Suppl): p. 57s-62s.

91 Antoniou, G.A., et al., *Clinical applications of robotic technology in vascular and endovascular surgery.* J Vasc Surg, 2011. 53(2): p. 493-9.

92 Evans, W.E., *Long-term evaluation of the celiac band syndrome.* Surgery, 1974. 76(6): p. 867-71.

93 Criado, E., *Transthoracic Median Arcuate Ligament Release and Mesenteric Revascularization.* Ann Vasc Surg, 2016. 33: p. 232-6.

94 Allen, R.C., et al., *Mesenteric angioplasty in the treatment of chronic intestinal ischemia.* J Vasc Surg, 1996. 24(3): p. 415-21; discussion 421-3.

95   Geelkerken, R.H. and J.H. van Bockel, *Mesenteric vascular disease: a review of diagnostic methods and therapies.* Cardiovasc Surg, 1995. **3**(3): p. 247-60.

96   Spies, J.B., et al., *Standards for interventional radiology. Standards of Practice Committee of the Society of Cardiovascular and Interventional Radiology.* J Vasc Interv Radiol, 1991. **2**(1): p. 59-65.

97   Dion, Y.M., et al., *Laparoscopic end-to-end aortobifemoral bypass with reimplantation of the inferior mesenteric artery. An experimental study.* Surg Endosc, 1999. **13**(5): p. 449-51.

98   Fajer, S., et al., *Laparoscopic repair of median arcuate ligament syndrome: a new approach.* J Am Coll Surg, 2014. **219**(6): p. e75-8.

99   van Petersen, A.S., et al., *Retroperitoneal endoscopic release in the management of celiac artery compression syndrome.* J Vasc Surg, 2009. **50**(1): p. 140-7.

100 Baccari, P., et al., *Celiac artery compression syndrome managed by laparoscopy.* J Vasc Surg, 2009. **50**(1): p. 134-9.

101 Roseborough, G.S., *Laparoscopic management of celiac artery compression syndrome.* J Vasc Surg, 2009. **50**(1): p. 124-33.

102 Tulloch, A.W., et al., *Laparoscopic versus open celiac ganglionectomy in patients with median arcuate ligament syndrome.* J Vasc Surg, 2010. **52**(5): p. 1283-9.

103 Aschenbach, R., S. Basche, and T.J. Vogl, *Compression of the celiac trunk caused by median arcuate ligament in children and adolescent subjects: evaluation with contrast-enhanced MR angiography and comparison with Doppler US evaluation.* J Vasc Interv Radiol, 2011. **22**(4): p. 556-61.

104 Berard, X., et al., *Laparoscopic surgery for coeliac artery compression syndrome: current management and technical aspects.* Eur J Vasc Endovasc Surg, 2012. **43**(1): p. 38-42.

105 Nguyen, T., et al., *Laparoscopic management of the median arcuate ligament syndrome.* ANZ J Surg, 2012. **82**(4): p. 265-8.

106 Do, M.V., et al., *Laparoscopic versus robot-assisted surgery for median arcuate ligament syndrome.* Surg Endosc, 2013. **27**(11): p. 4060-6.

107 El-Hayek, K.M., et al., *Laparoscopic median arcuate ligament release: are we improving symptoms?* J Am Coll Surg, 2013. **216**(2): p. 272-9.

108 Joyce, D.D., et al., *Pediatric median arcuate ligament syndrome: surgical outcomes and quality of life.* J Laparoendosc Adv Surg Tech A, 2014. **24**(2): p. 104-10.

109 Nasr, L.A., et al., *Median Arcuate Ligament Syndrome: A Single-Center Experience with 23 Patients.* Cardiovasc Intervent Radiol, 2017. **40**(5): p. 664-670.

110 Notani, N., et al., *Acute celiac artery compression syndrome after extensive correction of sagittal balance on an adult spinal deformity.* Eur Spine J, 2016.

111 Lamba, R., et al., *Multidetector CT of vascular compression syndromes in the abdomen and pelvis.* Radiographics, 2014. **34**(1): p. 93-115.

112 Brocker, J.A., J.L. Maher, and R.W. Smith, *True pancreaticoduodenal aneurysms with celiac stenosis or occlusion.* Am J Surg, 2012. **204**(5): p. 762-8.

113 Corey, M.R., et al., *The presentation and management of aneurysms of the pancreaticoduodenal arcade.* J Vasc Surg, 2016. **64**(6): p. 1734-1740.

114 Sakatani, A., et al., *Pancreaticoduodenal artery aneurysm associated with coeliac artery occlusion from an aortic intramural hematoma.* World J Gastroenterol, 2016. **22**(16): p. 4259-63.

115 Paty, P.S., et al., *Aneurysms of the pancreaticoduodenal artery.* J Vasc Surg, 1996. **23**(4): p. 710-3.

116 Quandalle, P., et al., *Pancreaticoduodenal artery aneurysms associated with celiac axis stenosis: report of two cases and review of the literature.* Ann Vasc Surg, 1990. **4**(6): p. 540-5.

117 Kalva, S.P., et al., *Inferior Pancreaticoduodenal Artery Aneurysms in Association with Celiac Axis Stenosis or Occlusion.* European Journal of Vascular and Endovascular Surgery. **33**(6): p. 670-675.

118 Bühler, A., *Uber eine anastomose zwischen den stämmen der Art. Coeliaca und der Art. Mesenterica Superior.* Morpholog Jahrb, 1904. **32**: p. 185-188.

119 Tandler, J., *Uber die varietaten der arteria coeliaca und deren entwicklung.* Anat Hefte, 1904. **25**: p. 472-500.

120 McNulty, J.G., et al., *Surgical and radiological significance of variants of Buhler's anastomotic artery: a report of three cases.* Surg Radiol Anat, 2001. **23**(4): p. 277-80.

121 Saad, W.E., et al., *Arc of buhler: incidence and diameter in asymptomatic individuals.* Vasc Endovascular Surg, 2005. **39**(4): p. 347-9.

122 Grabbe, E. and E. Bucheler, *[Buhler's anastomosis (author's transl)].* Rofo, 1980. **132**(5): p. 541-6.

123 Sugihara, F., et al., *Successful Coil Embolization of an Aneurysm in the Arc of Buhler.* J Nippon Med Sch, 2016. **83**(5): p. 196-198.